THOUGHTS -> FEELINGS -> ACTIONS

ENERGY DRIVEN AGILE TRANSFORMATION WORKBOOK

A SUPPORT GUIDE FOR INDIVIDUALS AND ORGANIZATION UNDERGOING AN AGILE IMPLEMENTATION

CHRISTOPHER A. LEWIS

COPYRIGHT © 2019 ALBERT CHRISTOPHER SOLUTIONS. ALL RIGHTS RESERVED.

PUBLISHED BY HILL & WEISS PUBLISHING, MIAMI FL.

NO PART OF THIS PUBLICATION MAY BE SHARED, DISTRIBUTED, OR USED FOR ANY PURPOSE OTHER THAN INTENDED, WHICH IS FOR THE USE WITHIN THE PARAMETERS OF A COACHING RELATIONSHIP. NOTHING IN THIS WORKBOOK MAY BE USED FOR WORKSHOPS OR TRAININGS WITHOUT THE WRITTEN CONSENT OF CHRISTOPHER LEWIS AND ALBERT CHRISTOPHER SOLUTIONS. YOUR USE OF THIS WORKBOOK ESTABLISHES ACCEPTANCE OF THESE TERMS.

TABLE OF CONTENTS

A NOTE FROM THE AUTHOR — 4

OVERVIEW AND PERSONAL GOALS — 5

ENERGY LEVELS AND ENERGETIC PROFILE — 9

ENERGETIC PROFILE COMBINATIONS — 20

PERSONAL VALUES AND EFFECTIVE COMMUNICATION — 28

WHAT ARE YOUR THOUGHTS? — 39

WHAT ARE YOUR FEELINGS? — 48

WHAT ARE YOUR ACTIONS? — 57

REFLECTION — 66

A NOTE FROM THE AUTHOR

MILLIONS OF BOOKS ARE AUTHORED TO TELL YOU WHAT TO DO OR TO SHARE PERSONAL THOUGHTS. THIS BOOK IS THE COMPLETE ANTITHESIS OF THOSE PUBLICATIONS AND IS ABOUT YOU DISCOVERING WHAT YOU WANT DURING AN AGILE TRANSFORMATION. THIS IS NOT A COOKIE CUTTER AGILE BOOK THAT WILL DETAIL HOW TO IMPROVE COMMON PRACTICES OF STAND UPS OR RETROSPECTIVES. THIS IS AN ENERGY DRIVEN **WORK**BOOK TO HELP GUIDE THE PROCESS CHANGE AND MORE SIGNIFICANTLY, THE PERCEPTION CHANGE NEEDED DURING AN AGILE IMPLEMENTATION. THIS WILL BE WORK.

IF YOU ARE UTILIZING THIS WORKBOOK, CHANCES ARE YOU ARE ABOUT TO BEGIN AN AGILE TRANSFORMATION, CURRENTLY WITHIN AN AGILE TRANSFORMATION, OR HAVE GONE THROUGH A "FAILED" AGILE TRANSFORMATION. I WROTE THIS BOOK FOR YOU. I WAS IN YOUR POSITION AND I USED MY FRUSTRATION TO CREATE THIS WORKBOOK SO THAT YOU, YOUR COWORKERS, AND YOUR ORGANIZATION CAN UNDERSTAND THAT AN AGILE TRANSFORMATION IS ABOUT MORE THAN PROCESSES AND RITUALS. IT IS TRULY ABOUT PEOPLE AND HOW THEIR ENERGY CAN SHIFT FROM A PLACE OF CONTROL TO A PLACE OF CREATIVITY.

WITHIN OUR PROFESSIONAL LIVES THERE ARE MANY INTERNAL AND EXTERNAL OBSTACLES THAT GET IN OUR WAY ALONG THE ROAD TO WHERE WE WOULD LIKE TO BE. THIS BOOK, WILL BE THE ONE THING THAT WILL NOT BE IN YOUR WAY. ENERGY DRIVEN AGILE TRANSFORMATION WORKBOOK WILL SIMPLY PROVIDE INFORMATION AND QUESTIONS, THAT YOU CAN CHOOSE TO IGNORE OR ACKNOWLEDGE ON YOUR AGILE JOURNEY, WITHOUT JUDGMENT.

LET'S BEGIN!

CHRISTOPHER A. LEWIS
AUTHOR AND COACH

OVERVIEW AND PERSONAL GOALS

BY THE END OF THIS SECTION, PEOPLE WITHIN THE ORGANIZATION WILL BE ABLE TO IDENTIFY THE TOP THREE THINGS THAT THEY WOULD LIKE TO ACCOMPLISH BY COMPLETING THE WORKBOOK.

WELCOME TO ENERGY DRIVEN AGILE. THIS WORKBOOK STEMS FROM CORE ENERGY COACHING ESTABLISHED BY BRUCE D. SCHNEIDER. HE IS THE FOUNDER OF INSTITUTE FOR PROFESSIONAL EXCELLENCE IN COACHING, AN INTERNATIONAL COACHING FEDERATION (ICF) ACCREDITED COACHING PROGRAM THAT SPECIALIZES IN IN HOW PEOPLE RESPOND IN STRESSFUL AND FAVORABLE SITUATIONS. THEIR RESPONSE CAN EITHER LEAD TO SUSTAINABLE GROWTH OR LEAD TO DESTRUCTIVE TENDENCIES WITHIN AN AGILE TRANSFORMATION. THIS IS THE FOUNDATION AS YOU MOVE FORWARD WITHIN THIS WORKBOOK.

TO GET THE MOST FROM THIS WORKBOOK:

- BE OPEN MINDED AND DO NOT TO JUDGE THE COACH, QUESTIONS, OR YOUR ANSWERS. GO WITH THE FLOW.
- TAKE TIME TO THINK ABOUT YOUR ANSWERS, SOME QUESTIONS MAY NEED TO BE SKIPPED AND REVISITED IF THE FULL ANSWER DOES NOT IMMEDIATELY SURFACE. THIS IS MORE THAN OKAY, JUST BE MINDFUL OF AVOIDANCE VERSUS IN DEPTH THINKING.
- SEE CHALLENGING QUESTIONS AND PEOPLE AS AN ADVANTAGE DURING THIS PROCESS.
- EACH MOMENT DESCRIBES WHO YOU ARE, AND GIVES YOU THE OPPORTUNITY TO DECIDE IF THAT IS WHO YOU WANT TO BE. IF YOU WANT TO ACCOMPLISH YOUR GOALS BY COMPLETING THIS WORKBOOK, STAY IN THE MOMENT.

LIST YOUR 3 MAIN GOALS THAT YOU WOULD LIKE TO ACCOMPLISH BY COMPLETING THIS WORKBOOK.

ENERGY DRIVEN AGILE: DRIVE PRINCIPLES

1. THE PAST IS NEVER THE PROBLEM

2. RESPONSIBILITY AND ACCOUNTABILITY ARE VEHICLES FOR REWARD

3. WHAT YOU ARE WILLING TO GIVE MAY BE THE MOST SIGNIFICANT THING, WHEN TRYING TO OBTAIN WHAT YOU WANT

4. WHEN YOU CHANGE WHO YOU ARE IN A RELATIONSHIP, THE RELATIONSHIP ITSELF CHANGES

5. YOUR TRUE DESIRES, NOT YOUR INTERPRETATIONS, GUIDE YOUR ACTIONS

6. FOCUS ON COMMITMENTS OVER RESULTS

7. THE TRUTH IS GREATER THAN YOUR TRUTH

ENERGY LEVELS AND ENERGETIC PROFILES

BY THE END OF THIS SECTION, PEOPLE WITHIN THE ORGANIZATION WILL BE ABLE TO LIST AND DEFINE THE CORE THOUGHTS, FEELINGS, AND ACTIONS OF EACH ENERGY LEVEL.

WHAT IS AN ENERGY LEVEL?

THE LENS IN WHICH AN INDIVIDUAL CHOOSES TO VIEW A GIVEN SITUATION AND THE HIGHER THE RESOLUTION (ENERGY LEVEL) THE GREATER THE POTENTIAL TO SEE AN OPPORTUNITY TO CREATE SOMETHING THAT DID NOT EXIST PRIOR.

MAIN THEMES

CORE THOUGHT
IDEA OR BELIEF THAT I HOLD ONTO

CORE FEELING
EMOTION I INTERNALIZE, INFLUENCED BY MY THOUGHTS

CORE ACTION
WHAT I DO (BEHAVIOR), INFLUENCED BY MY EMOTIONS

LEVEL 1
"THE VICTIM"

CORE THOUGHT
I AM HELPLESS TO EVENTS AND THOUGHTS THAT HOLD ME BACK

CORE FEELING
I AM INDIFFERENT OR SAD

CORE ACTION
I DO NOT TAKE ANY ACTION OR INITIATIVE

LEVEL 2

"THE FIGHTER"

CORE THOUGHT

I JUDGE EVERYTHING AS RIGHT OR WRONG

CORE FEELING

I AM ANGRY AND IRRITABLE

CORE ACTION

I ENGAGE IN RESISTANCE AND RIDICULE

LEVEL 3

"THE RATIONALIZER"

CORE THOUGHT

I AM ACCOUNTABLE FOR MY OWN EXPERIENCES AND CHOICES

CORE FEELING

I AM FORGIVING

CORE ACTION

I SEEK TEAMWORK AND COOPERATION

LEVEL 4
"THE CAREGIVER"

CORE THOUGHT
I AM CONCERNED ABOUT OTHERS AND MYSELF

CORE FEELING
I AM COMPASSIONATE AND SYMPATHETIC

CORE ACTION
I PROVIDE HELP OR ASSITANCE

LEVEL 5
"THE OPPORTUNIST"

CORE THOUGHT
I DO NOT JUDGE AND SEEK THE WIN/WIN

CORE FEELING
I AM CALM AND SATISFIED

CORE ACTION
I FIND THE OPPORTUNITY AND VALUE IN ALL THINGS

LEVEL 6
"THE VISIONARY"

CORE THOUGHT
I THINK WE ARE ALL ONE, NO JUDGMENT

CORE FEELING
I AM JOYOUS

CORE ACTION
I PROVIDE WISDOM STEMMING FROM MY INTUITION

LEVEL 7
"THE CREATOR"

CORE THOUGHT
I HAVE NO FEAR OR JUDGMENT

CORE FEELING
I AM INTENSE AND BLISSFUL ABOUT ALL THINGS AND EXPERIENCES

CORE ACTION
I CREATE NEW IDEAS AND THOUGHTS

NOTE, THE NEXT SECTION'S OPTIMAL UNDERSTANDING OCCURS WHEN AN ENERGY LEADERSHIP ASSESSMENT HAS BEEN COMPLETED. WITHOUT THIS ASSESSMENT YOU WILL NOT TRULY KNOW YOUR AVERAGE RESONATING LEVEL OF ENERGY.

TO SCHEDULE AN ELI ASSESSMENT AND DEBRIEF, PLEASE EMAIL ACS.CHRISLEWIS@GMAIL.COM OR VISIT WWW.ENERGYDRIVENAGILE.COM.

ENERGETIC PROFILES COMBINATIONS

BY THE END OF THIS SECTION, PEOPLE WITHIN THE ORGANIZATION WILL BE ABLE TO GIVE ORIGINAL EXAMPLES OF CURRENT WORKPLACE INTERACTIONS BASED ON THE ENERGY LEVEL COMBINATIONS AND DISCUSS WHICH COMBINATIONS ARE OPTIMAL FOR A SUCCESSFUL AGILE TRANSFORMATION AND QUALITY WORK-LIFE.

FIND YOUR LEVEL 1 COMBINATIONS

AFTER THE ENERGY LEADERSHIP ASSESSMENT (ELI) AND THE AVERAGE RESONATING LEVEL OF ENERGY (ARL) IS PRESENT. REVIEW HOW YOUR ARL RELATES TO ANOTHER PERSON OR GROUP.

LEVEL 1 - LEVEL 1 (VICTIM/VICTIM)

STRENGTH: BOTH WILL RECEIVE ATTENTION AND SYMPATHY FROM OTHERS
CHALLENGE: BOTH AVOID TAKING ACTION TO ADDRESS WHAT IS BOTHERSOME

LEVEL 1 - LEVEL 2 (VICTIM/FIGHTER)

STRENGTH: THINGS GET ACCOMPLISHED, ALBEIT BY FORCE
CHALLENGE: ACTION MOTIVATED BY FEAR AND BLAME

LEVEL 1 - LEVEL 3 (VICTIM/RATIONALIZER)

STRENGTH: THINGS GET ACCOMPLISHED, ALBEIT BY FORGIVENESS
CHALLENGE: LEVEL 3 COPES WITH LEVEL 1 TO BE PRODUCTIVE

LEVEL 1 - LEVEL 4 (VICTIM/CAREGIVER)

STRENGTH: LEVEL 4 WILL PROTECT LEVEL 1 FROM HARM
CHALLENGE: LEVEL 4 EXPERIENCES "BURN OUT" OR RESENTMENT

LEVEL 1 - LEVEL 5 (VICTIM/OPPORTUNIST)

STRENGTH: LEVEL 5 HELPS LEVEL 1 SEE OPTIONS
CHALLENGE: LEVEL 1 FEELS FEELINGS DO NOT MATTER; LEVEL 5 FEELS MISREAD

LEVEL 1 - LEVEL 6 (VICTIM/VISIONARY)

STRENGTH: LEVEL 6 IS EMPATHETIC TOWARDS LEVEL 1
CHALLENGE: LEVEL 6 IS PERCEIVED AS OUT OF TOUCH AND FEELS MISREAD

FIND YOUR LEVEL 2 COMBINATIONS

AFTER THE ENERGY LEADERSHIP ASSESSMENT (ELI) AND THE AVERAGE RESONATING LEVEL OF ENERGY (ARL) IS PRESENT. REVIEW HOW YOUR ARL RELATES TO ANOTHER PERSON OR GROUP.

LEVEL 2 - LEVEL 1 (FIGHTER/VICTIM)
STRENGTH: THINGS GET ACCOMPLISHED, ALBEIT BY FORCE
CHALLENGE: ACTION IS MOTIVATED BY FEAR AND BLAME

LEVEL 2 - LEVEL 2 (FIGHTER/FIGHTER)
STRENGTH: BOTH TAKE ACTION TO PROTECT THEMSELVES
CHALLENGE: BOTH PLACE BLAME ON OTHERS AND FIGHT TO GET WHAT THEY WANT

LEVEL 2 - LEVEL 3 (FIGHTER/RATIONALIZER)
STRENGTH: THINGS GET ACCOMPLISHED, ALBEIT BY COOPERATION
CHALLENGE: LEVEL 3 TOLERATES THINGS TO RECEIVE COOPERATION

LEVEL 2 - LEVEL 4 (FIGHTER/CAREGIVER)
STRENGTH: LEVEL 4 WILL ENSURE LEVEL 2 RECEIVES WHAT THEY DESIRE
CHALLENGE: LEVEL 4 EXPERIENCES "BURN OUT" OR RESENTMENT

LEVEL 2 - LEVEL 5 (FIGHTER/OPPORTUNIST)
STRENGTH: LEVEL 5 WILL HELP LEVEL 2 SEE THE VALUE IN PERSONAL ACCOUNTABILITY
CHALLENGE: LEVEL 2 IS FRUSTRATED FORCE IS INEFFECTIVE; LEVEL 5 FEELS MISREAD

LEVEL 2 - LEVEL 6 (FIGHTER/VISIONARY)
STRENGTH: LEVEL 6 IS EMPATHETIC TOWARDS LEVEL 2
CHALLENGE: LEVEL 6 IS PERCEIVED AS OUT OF TOUCH AND FEELS MISREAD

FIND YOUR LEVEL 3 COMBINATIONS

AFTER THE ENERGY LEADERSHIP ASSESSMENT (ELI) AND THE AVERAGE RESONATING LEVEL OF ENERGY (ARL) IS PRESENT. REVIEW HOW YOUR ARL RELATES TO ANOTHER PERSON OR GROUP.

LEVEL 3 - LEVEL 1 (RATIONALIZER/VICTIM)
STRENGTH: THINGS GET ACCOMPLISHED, ALBEIT BY FORGIVENESS
CHALLENGE: LEVEL 3 COPES WITH LEVEL 1 TO CONTINUE TO BE PRODUCTIVE

LEVEL 3 - LEVEL 2 (RATIONALIZER/FIGHTER)
STRENGTH: THINGS GET ACCOMPLISHED, ALBEIT BY COOPERATION
CHALLENGE: LEVEL 3 TOLERATES LEVEL 2 TO RECEIVE COOPERATION

LEVEL 3 - LEVEL 3 (RATIONALIZERX2)
STRENGTH: BOTH ARE DRIVEN BY WHAT THEY WANT TO ACHIEVE
CHALLENGE: BOTH ARE HIGHLY FOCUSED ON ACHIEVING THEIR OWN PERSONAL INTERESTS

LEVEL 3 - LEVEL 4 (RATIONALIZER/CAREGIVER)
STRENGTH: LEVEL 4 WILL ENSURE LEVEL 3 MEETS THEIR GOALS
CHALLENGE: LEVEL 4 EXPERIENCES "BURN OUT" OR RESENTMENT

LEVEL 3 - LEVEL 5 (RATIONALIZER/OPP.)
STRENGTH: LEVEL 5 WILL AID LEVEL 3 TO SEE THINGS THAT THEY MAY TOLERATE
CHALLENGE: LEVEL 5 FEELS MISREAD

LEVEL 3 - LEVEL 6 (RATIONALIZER/VISIONARY)
STRENGTH: LEVEL 6 WILL AID LEVEL 3 SEE THE VALUE IN EVERYONE'S GOALS
CHALLENGE: LEVEL 6 FEELS MISREAD

FIND YOUR LEVEL 4 COMBINATIONS

AFTER THE ENERGY LEADERSHIP ASSESSMENT (ELI) AND THE AVERAGE RESONATING LEVEL OF ENERGY (ARL) IS PRESENT. REVIEW HOW YOUR ARL RELATES TO ANOTHER PERSON OR GROUP.

LEVEL 4 - LEVEL 1 (CAREGIVER/VICTIM)
STRENGTH: LEVEL 4 WILL PROTECT LEVEL 1 FROM HARM
CHALLENGE: LEVEL 4 MAY EXPERIENCE "BURN OUT" OR RESENTMENT

LEVEL 4 - LEVEL 2 (CAREGIVER/FIGHTER)
STRENGTH: LEVEL 4 WILL ENSURE LEVEL 2 RECEIVES WHAT THEY DESIRE
CHALLENGE: LEVEL 4 MAY EXPERIENCE "BURN OUT" OR RESENTMENT

LEVEL 4 - LEVEL 3 (CAREGIVER/RATIONALIZER)
STRENGTH: LEVEL 4 WILL ENSURE LEVEL 3 MEETS THEIR GOALS
CHALLENGE: LEVEL 4 MAY EXPERIENCE "BURN OUT" OR RESENTMENT

LEVEL 4 - LEVEL 4 (CAREGIVER/CAREGIVER)
STRENGTH: BOTH ARE FULLY DEDICATED TO HELPING OTHERS (INCLUDING EACH OTHER) WIN
CHALLENGE: BOTH TAKE ON OTHER PEOPLE'S ISSUES AS IF THEY WERE THEIR OWN

LEVEL 4 - LEVEL 5 (CAREGIVER/OPPORTUNIST)
STRENGTH: LEVEL 5 WILL AID LEVEL 4 IN CREATING WIN/WIN SITUATIONS
CHALLENGE: LEVEL 5 SEES OPPORTUNITIES FOR WIN/WIN SITUATIONS THAT LEVEL 4 MAY NOT AND FEELS MISREAD

LEVEL 4 - LEVEL 6 (CAREGIVER/VISIONARY)
STRENGTH: LEVEL 6 WILL AID LEVEL 4 IN REMEMBERING BY CARING WE ALL WIN
CHALLENGE: LEVEL 6 IS PERCEIVED AS OUT OF TOUCH AND FEELS MISREAD

FIND YOUR LEVEL 5 COMBINATIONS

AFTER THE ENERGY LEADERSHIP ASSESSMENT (ELI) AND THE AVERAGE RESONATING LEVEL OF ENERGY (ARL) IS PRESENT. REVIEW HOW YOUR ARL RELATES TO ANOTHER PERSON OR GROUP.

LEVEL 5 - LEVEL 1 (OPPORTUNIST/VICTIM)
STRENGTH: LEVEL 5 HELPS LEVEL 1 SEE OPTIONS
CHALLENGE: LEVEL 1 FEELS THEIR FEELINGS DO NOT MATTER; LEVEL 5 FEELS MISREAD

LEVEL 5 - LEVEL 2 (OPPORTUNIST/FIGHTER)
STRENGTH: LEVEL 5 WILL HELP LEVEL 2 SEE THEIR PERSONAL RESPONSIBILITY FOR THEIR DESIRES
CHALLENGE: LEVEL 2 IS ANGRY THAT FORCE IS NOT EFFECTIVE; LEVEL 5 FEELS MISREAD

LEVEL 5 - LEVEL 3 (OPP./RATIONALIZER)
STRENGTH: LEVEL 5 WILL AID LEVEL 3 TO SEE THINGS THAT THEY MAY TOLERATE
CHALLENGE: LEVEL 5 FEELS MISREAD

LEVEL 5 - LEVEL 4 (OPPORTUNIST/CAREGIVER)
STRENGTH: LEVEL 5 WILL AID LEVEL 4 IN CREATING A WIN/WIN SITUATION
CHALLENGE: LEVEL 5 SEES USEFUL PERSONAL BOUNDARIES, THAT LEVEL 4 DOES NOT SEE

LEVEL 5 - LEVEL 5 (OPPORTUNISTX2)
STRENGTH: BOTH SEE OPPORTNITIES IN ALL THINGS AND TAKE NOTHING PERSONALLY
CHALLENGE: BOTH ARE PERCEIVED AS OUT OF TOUCH AND OVERLY EMBRACE RISK

LEVEL 5 - LEVEL 6 (OPPORTUNIST/VISIONARY)
STRENGTH: BOTH HAVE GREAT IDEAS AND KNOW HOW TO IMPLEMENT THEM
CHALLENGE: BOTH ARE PERCEIVED AS OUT OF TOUCH AND OVERLY EMBRACE RISK

FIND YOUR LEVEL 6 COMBINATIONS

AFTER THE ENERGY LEADERSHIP ASSESSMENT (ELI) AND THE AVERAGE RESONATING LEVEL OF ENERGY (ARL) IS PRESENT. REVIEW HOW YOUR ARL RELATES TO ANOTHER PERSON OR GROUP.

LEVEL 6 - LEVEL 1 (VISIONARY/VICTIM)
STRENGTH: LEVEL 6 IS EMPATHETIC TOWARDS LEVEL 1
CHALLENGE: LEVEL 6 IS PERCEIVED AS OUT OF TOUCH AND FEELS MISREAD

LEVEL 6 - LEVEL 2 (VISIONARY/FIGHTER)
STRENGTH: LEVEL 6 IS EMPATHETIC TOWARDS LEVEL 2
CHALLENGE: LEVEL 6 IS PERCEIVED AS OUT OF TOUCH AND FEELS MISREAD

LEVEL 6 - LEVEL 3 (VISIONARY/RATIONALIZER)
STRENGTH: LEVEL 6 WILL AID LEVEL 3 SEE THE VALUE IN EVERYONE MEETING THEIR GOALS
CHALLENGE: LEVEL 6 FEELS MISREAD

LEVEL 6 - LEVEL 4 (VISIONARY/CAREGIVER)
STRENGTH: LEVEL 6 WILL AID LEVEL 4 IN REMEMBERING BY CARING WE ALL WIN
CHALLENGE: LEVEL 6 IS PERCEIVED AS OUT OF TOUCH AND FEELS MISREAD

LEVEL 6 - LEVEL 5 (VISIONARY/OPPORTUNIST)
STRENGTH: BOTH HAVE GREAT IDEAS AND KNOW HOW TO IMPLEMENT THEM
CHALLENGE: BOTH ARE PERCEIVED AS OUT OF TOUCH AND OVERLY EMBRACE RISK

LEVEL 6 - LEVEL 6 (VISIONARY/VISIONARY)
STRENGTH: BOTH HAVE A DEEP CONNECTIONS WITH PEOPLE, WITHOUT JUDGMENT
CHALLENGE: BOTH ARE PERCEIVED AS OUT OF TOUCH AND OVERLY EMBRACE RISK

WHAT STANDS OUT FOR YOU AFTER REVIEWING THE ENERGETIC PROFILE COMBINATIONS?

WHERE IN YOUR ORGANIZATION DO YOU ENCOUNTER LEVEL 1 OR LEVEL 2 ENERGY THE MOST?

WHERE IN YOUR ORGANIZATION DO YOU ENCOUNTER LEVELS 4 AND UP THE MOST?

WHAT ARE YOUR VALUES AND HOW DO THEY INFLUENCE EFFECTIVE COMM. FOR YOU?

BY THE END OF THIS SECTION, PEOPLE WILL BE ABLE TO DEMONSTRATE THEIR, NEWLY DISCOVERED, PERSONAL VALUES DAILY. THIS MAY LEAD TO PEOPLE BECOMING MORE AUTHENTIC WITHIN THE ORGANIZATION ENABLING A COLLABORATIVE CULTURE ABLE TO PERFORM THE EFFECTIVE COMMUNICATION TECHNIQUES TO IMPROVE DAILY COMMUNICATION.

LIST OF 50 PERSONAL VALUES

1. CIRCLE ALL OF THE VALUES THAT STAND OUT FOR YOU THE MOST.
2. PLACE A STAR OR CHECK NEXT TO YOUR TOP 5 STAND OUT VALUES.

1. ACCOUNTABILITY
2. ADVENTUROUSNESS
3. BALANCE
4. CALMNESS
5. CHEERFULNESS
6. CREATIVITY
7. COMMUNITY
8. COMPASSION
9. COMPETITIVENESS
10. COURAGE
11. COURTEOUSNESS
12. CURIOSITY
13. DECISIVENESS
14. DETERMINATION
15. DISCIPLINE
16. EFFECTIVENESS
17. EFFICIENCY
18. ENJOYMENT
19. ENTHUSIASM
20. EMPATHY
21. EXPLORATION
22. FAIRNESS
23. FREEDOM
24. GENEROSITY
25. HEALTH
26. HONESTY
27. HUMILITY
28. INTELLIGENCE
29. INQUISITIVENESS
30. JOY
31. LOYALTY
32. MASTERY
33. OPENNESS
34. OBEDIENCE
35. PERFECTION
36. PREPAREDNESS
37. RELIABILITY
38. SELF-CONTROL
39. SELFLESSNESS
40. SIMPLICITY
41. SPEED
42. SPONTANEITY
43. STABILITY
44. STRENGTH
45. SUCCESS
46. TEAMWORK
47. TIMELINESS
48. TRUSTWORTHINESS
49. TRUTH SEEKING
50. UNIQUENESS

WHERE DO YOU BELIEVE YOUR TOP VALUES 3 COME FROM?

SINCE YOU HAVE REVIEWED THE ENERGY LEVELS, WHAT LEVEL OF ENERGY DOES EACH OF YOUR TOP 3 VALUES ORIGINATE FROM?

FROM A SCALE FROM 1-10, 1 BEING NEVER AND 10 BEING ALWAYS. HOW LIKELY ARE YOU LIVING YOUR TOP 3 VALUES WITHIN YOUR DAILY LIFE WITHIN YOUR ORGANIZATION?

ACKNOWLEDGMENT

ACKNOWLEDGING IS ONE OF THE MOST POWERFUL PARTS OF COMMUNICATION. WHEN ACKNOWLEDGING SOMEONE YOU LET THEM KNOW THAT YOU HAVE REALLY LISTENED AND MIRROR BACK OR PARAPHRASE WHAT THEY HAVE SAID.

COMMON ACKNOWLEDGING APPROACHES:
WHAT YOU ARE SAYING IS…
LET ME SEE IF I GET THIS RIGHT…

VALIDATION

VALIDATING IS ALSO ONE OF THE MOST POWERFUL PARTS OF COMMUNICATION. WE ALL HAVE FEELINGS AND WHEN WE VALIDATE ANOTHER PERSON WE ARE LETTING THEM KNOW THEY HAVE A RIGHT TO FEEL THAT WAY. VALIDATION IS NOT A JUDGMENT OF RIGHT OR WRONG NOR IS IT AN AGREEMENT. IT IS AN ACKNOWLEDGMENT OF THEIR PERSPECTIVE. THIS IS AN ESSENTIAL PART OF EFFECTIVE COMMUNICATION.

COMMON VALIDATION APPROACHES:
WHO WOULDN'T FEEL THAT WAY WHEN X HAPPENS.
BASED ON YOUR VALUES, IT MAKES SENSE YOU FEEL THAT WAY.

CLARIFICATION

CLARIFYING IS MAKING SURE THAT, WHEN SPEAKING WITH SOMEONE ELSE, WE KNOW WHAT A PERSON MEANS WHEN THEY USE CERTAIN WORDS OR PHRASES. IT IS ALSO SIGNIFICANT BECAUSE IT HELPS THE SPEAKER UNDERSTAND WHAT THEY MEAN WHEN THEY SAY CERTAIN WORDS OR PHRASES. THIS HELPS ENCOURAGE THE SPEAKER TO SPEAK MORE AND PROMOTES DEEPER UNDERSTANDING.

COMMON CLARIFYING QUESTIONS:
WHAT DO YOU MEAN WHEN YOU SAY...?
TELL ME WHAT IS "X" ABOUT IT?

WHAT WOULD BE THE BENEFIT, IF ANY, OF YOU IMPLEMENTING THESE COMMUNICATION TOOLS MORE OFTEN IN YOUR DAILY COMMUNICATION WITHIN YOUR ORGANIZATION?

ENERGY BLOCKERS

INNER CRITIC
TELLS YOU ARE NOT GOOD ENOUGH, IN ONE WAY OR ANOTHER

ASSUMPTIONS
THINKING BECAUSE SOMETHING HAPPENED IN THE PAST, IT WILL HAPPEN AGAIN

INTERPRETATIONS
OPINION OR JUDGMENT THAT YOU CREATE ABOUT A SITUATION OR PERSON

LIMITING BELIEFS
SOMETHING YOU ACCEPT ABOUT LIFE OR YOURSELF THAT LIMITS YOU IN SOME WAY

WHEN DO YOU TELL YOURSELF THAT YOU ARE NOT GOOD ENOUGH WITHIN YOUR ORGANIZATION?

WHAT LEADS YOU TO BELIEVE THAT YOU ARE "LESS THAN" IN THIS AREA WITHIN YOUR ORGANIZATION?

WHAT DO YOU THINK WOULD BE THE BENEFIT FOR YOU, IF YOU WANTED TO, IF YOU QUIETED THIS INNER CRITICISM?

DRIVE PRINCIPLE #1

THE PAST IS NEVER THE PROBLEM

THIS PRINCIPLE WHILE ACKNOWLEDGING THE PAST ENCOURAGES FORWARD, SOLUTION FOCUSED THINKING TO ACHIEVE WHAT YOU TRULY DESIRE. THERE IS NOTHING THAT CAN BE DONE TO CHANGE THE PAST, THE PRESENT MOMENT IS THE ONLY OPTION TO MOVE THE POSSIBILITY FORWARD.

WHEN WAS A TIME WHEN SOMETHING DID NOT GO AS YOU PLANNED WITHIN YOUR ORGANIZATION?

WHAT LEADS YOU TO BELIEVE THAT IF THE SAME SITUATION ARISES, THE RESULT WILL BE THE SAME?

WHAT DID YOU LEARN FROM THE FIRST EXPERIENCE, THAT MAY HELP OR HELPED YOU NOT ENCOUNTER THE SAME RESULT?

WHEN WAS A TIME WITHIN YOUR ORGANIZATION, WHEN SOMEONE DID SOMETHING AND IT "PUSHED YOUR BUTTONS"?

WHAT DO YOU BELIEVE THEIR ACTIONS SAID ABOUT YOU, DESPITE THAT PERSON NOT EXPLICITLY STATING ANYTHING ABOUT YOU?

WHAT BOTHERED YOU ABOUT THIS?

WHAT IS A BELIEF THAT YOU HOLD ONTO IN ORDER TO GAIN WHAT YOU DESIRE WITHIN YOUR ORGANIZATION?

WHERE DID THIS IDEA COME FROM?

WHAT LEADS YOU TO BELIEVE THIS IS THE TRUTH (NOT DEBATABLE)?

WHAT ARE YOUR THOUGHTS?

BY THE END OF THIS SECTION, PEOPLE WITHIN THE ORGANIZATION WILL BE ABLE TO BREAK DOWN WHAT THEIR THOUGHTS ARE ABOUT SPECIFIC AREAS OF THE AGILE TRANSFORMATION.

WHAT IS THE VALUE OF THE AGILE TRANSFORMATION FOR YOU?

WHAT DO YOU WANT TO EXPERIENCE DURING THE TRANSFORMATION?

WHAT IS STOPPING YOU?

DRIVE PRINCIPLE #2

RESPONSIBILITY AND ACCOUNTABILITY ARE VEHICLES FOR REWARD

THIS PRINCIPLE IS A CALL TO ACTION FOR YOU TO FEEL COMFORTABLE BEING SOMEONE WHO IS IN CHARGE OF THE AMOUNT OF EFFORT THAT IS EXERTED TO ACCOMPLISH SOMETHING THAT IS HIGHLY DESIRED. ACCEPT THAT YOU HAVE THE POWER TO MAKE MODIFICATIONS THROUGHOUT THE PROCESS.

WHAT IS SIGNIFICANT ABOUT AGILE GOING BEYOND PILOT PROJECTS FOR YOU?

WHAT IS THE BEST THING THAT COULD HAPPEN?

HOW DOES THIS FIT INTO YOUR CAREER PLANS?

WHAT ORGANIZATIONAL CULTURE CHANGES DO YOU THINK WILL OCCUR WITH THIS AGILE TRANSFORMATION?

WHAT IS THE WORST THING THAT COULD HAPPEN CONCERNING THE CHANGE IN ORGANIZATIONAL VALUES FOR YOU?

WHERE DO YOU THINK THAT BELIEF COMES FROM?

WHAT IS SIGNIFICANT FOR YOU ABOUT THE ORGANIZATION INVESTING IN ITS STAFF DURING THIS AGILE TRANSFORMATION ?

HOW DO YOU DEFINE INVESTMENT?

WHERE DO YOU THINK THE BLIND SPOT IS CONCERNING TRAINING DURING THIS AGILE TRANSFORMATION?

WHAT IS SIGNIFICANT FOR YOU ABOUT DEFINING THE STRATEGY OF SCALING AGILE BEYOND PILOT PROJECTS?

WHAT ARE YOU UNWILLING TO TOLERATE DURING THE EXECUTION OF THE SCALED AGILE STRATEGY?

WHAT SPECIFICALLY ABOUT THAT INTOLERABLE "THING" HAS MEANING FOR YOU?

WHAT IS SIGNIFICANT FOR YOU ABOUT SENIOR LEADERSHIP SUPPORT FOR THE AGILE TRANSFORMATION?

WHAT DOES SUPPORT LOOK LIKE FOR YOU?

WHAT DO YOU THINK THE MAIN SUPPORT CHALLENGE WILL BE MOVING FORWARD?

WHAT IS SIGNIFICANT FOR YOU ABOUT BEING ABLE TO EXPERIMENT AND LEARN DURING THIS AGILE TRANSFORMATION?

WHAT ARE THE BENEFITS OF THIS FOR YOU?

WHAT IF THIS ABILITY TO EXPERIMENT AND LEARN DOES NOT EXIST?

WHAT ARE YOUR FEELINGS?

BY THE END OF THIS SECTION, PEOPLE WITHIN THE ORGANIZATION WILL BE ABLE TO CHOOSE WHAT THEY FEEL ABOUT SPECIFIC AREAS OF THE AGILE TRANSFORMATION.

IN ONE WORD, WHAT DO YOU FEEL WHEN YOU THINK OF THE VALUE OF THE AGILE TRANSFORMATION FOR YOU?

WHAT SPECIFICALLY IS DRIVING THIS FEELING?

WHAT WOULD YOUR LIFE LOOK LIKE 6 MONTHS FROM NOW IF THIS FEELING WAS IMPROVED?

IN ONE WORD, WHAT DO YOU FEEL WHEN YOU THINK ABOUT AGILE GOING BEYOND PILOT PROJECTS AND BECOMING THE ORGANIZATIONAL NORM?

WHAT SPECIFICALLY IS DRIVING THIS FEELING?

WHERE DO YOU THINK THIS BELIEF COMES FROM?

IN ONE WORD, WHAT DO YOU FEEL WHEN YOU THINK ABOUT THE CHANGE
IN ORGANIZATIONAL CULTURE FROM THE AGILE TRANSFORMATION ?

WHAT SPECIFICALLY IS DRIVING THIS FEELING?

WHERE WILL THIS FEELING LEAD TO?

IN ONE WORD, WHAT DO YOU FEEL WHEN YOU THINK ABOUT THE ORGANIZATION INVESTING IN STAFF MEMBERS DURING THE AGILE TRANSFORMATION?

WHAT SPECIFICALLY IS DRIVING THIS FEELING?

IF YOU WERE TO LOOK AT INVESTMENT ANOTHER WAY, WHAT DO YOU THINK MIGHT OCCUR?

DRIVE PRINCIPLE #3

WHAT YOU ARE WILLING TO GIVE MAY BE THE MOST SIGNIFICANT THING, WHEN TRYING TO OBTAIN WHAT YOU WANT

THIS PRINCIPLE IS URGING YOU TO ASK YOURSELF WHAT YOU CAN CONTRIBUTE TO YOUR OWN DESIRE. THIS PLACES THE POWER IN YOUR HANDS AND ENABLES YOU TO FOCUS ON A WIN/WIN MENTALITY.

IN ONE WORD, WHAT DO YOU FEEL WHEN YOU THINK ABOUT THE ORGANIZATION'S SCALED AGILE STRATEGY?

WHAT DO YOU REALLY MEAN BY THAT?

HOW CAN YOU STRETCH YOURSELF IN THIS AREA?

IN ONE WORD, WHAT DO YOU FEEL WHEN YOU THINK ABOUT SENIOR LEADERSHIP SUPPORT DURING THE AGILE TRANSFORMATION?

WHAT WILL YOU FEEL AFTER THE AGILE TRANSFORMATION IS COMPLETED?

IF YOU WANTED TO, HOW CAN YOU FEEL THAT WAY NOW?

IN ONE WORD, WHAT DO YOU FEEL WHEN YOU THINK ABOUT EXPERIMENTATION AND LEARNING WITHIN THE AGILE TRANSFORMATION?

WHAT DO YOU THINK THE MAIN CHALLENGE WILL BE?

WHAT IS THE BEST THING THAT COULD HAPPEN?

WHAT ARE YOUR ACTIONS?

BY THE END OF THIS SECTION, PEOPLE WITHIN THE ORGANIZATION WILL BE ABLE TO CREATE ACTION PLANS FOR EACH SPECIFIC AREA OF THE AGILE TRANSFORMATION TO GAIN WHAT THEY DESIRE OUT OF THE TRANSFORMATION.

BASED ON HOW YOU FEEL ABOUT THE VALUE OF AGILE, WHAT DO YOU WANT?

HOW WELL IS WHAT YOU ARE DOING GETTING YOU WHAT YOU WANT?

WHAT IS THE FIRST THING THAT NEEDS TO BE ADDRESSED TO IMPROVE OR ENHANCE GETTING WHAT YOU WANT FROM THIS AGILE TRANSFORMATION?

DRIVE PRINCIPLE #4

WHEN YOU CHANGE WHO YOU ARE IN A RELATIONSHIP, THE RELATIONSHIP ITSELF CHANGES

THIS PRINCIPLE IS ENCOURAGES YOU TO FOCUS ON WHAT YOU DO WANT AND NOT WHAT YOU DO NOT WANT. UNDERSTANDING WHO YOU ARE AND WHAT IS DRIVING YOUR BEHAVIOR IS ESSENTIAL TO GETTING WHAT YOU DESIRE OUT OF A RELATIONSHIP.

BASED ON HOW YOU FEEL ABOUT AGILE BEING THE NEW NORMAL WHAT IS YOUR NEXT STEP?

HOW WILL THAT BENEFIT YOU?

WHAT CAN POSSIBLY GET IN YOUR WAY?

BASED ON HOW YOU FEEL ABOUT THE IMMINENT CHANGE IN ORGANIZATIONAL CULTURE, WHAT 2-3 OPTIONS DO YOU HAVE?

IF THOSE DO NOT GET YOU WHAT YOU DESIRE, WHAT ELSE CAN YOU DO?

ON A SCALE FROM 1-10, 1 BEING THE LOWEST AND 10 BEING THE HIGHEST. HOW CONFIDENT ARE YOU IN YOUR NEXT STEP(S)?

BASED ON HOW YOU FEEL ABOUT THE ORGANIZATION INVESTING IN ITS STAFF, WHAT CAN YOU DO TO MAKE THAT INVESTMENT IN YOURSELF?

WHAT WOULD THAT INVESTMENT COST YOU?

WHAT ABOUT THE COST SCARES YOU?

BASED ON HOW YOU FEEL ABOUT THE ORGANIZATION'S SCALED AGILE STRATEGY, WHAT WILL YOU DO NEXT?

WHAT IS YOUR BACKUP PLAN?

WHAT DO YOU NEED TO GET THAT DONE?

BASED ON HOW YOU FEEL ABOUT SENIOR LEADERSHIP SUPPORT FOR THE AGILE TRANSFORMATION, WHAT WILL YOU DO NEXT?

WHAT DO YOU NEED TO GET THIS DONE?

WHAT CAN POTENTIALLY GET IN YOUR WAY?

BASED ON HOW YOU FEEL ABOUT EXPERIMENTATION AND LEARNING DURING THE AGILE PROCESS, WHAT DID YOU LEARN FROM THAT?

HOW CAN WHAT YOU LEARNED HELP YOU MOVING FORWARD?

WHAT IS THE NEXT STEP?

PERSONAL REFLECTION

BY THE END OF THIS SECTION, PEOPLE WITHIN THE ORGANIZATION WILL BE ABLE TO DETERMINE IF THERE ARE AREAS THAT THEY WOULD LIKE TO EXPLORE FURTHER BYWAY OF ONE ON ONE OR GROUP COACHING.

WHAT WAS YOUR BIGGEST TAKEAWAY FROM THIS WORKBOOK?

WHAT SURPRISED YOU MOST ABOUT THIS NEW DISCOVERY?

WHAT IS IN IT FOR YOU TO IMPROVE IN THE(SE) AREA(S) DURING THE AGILE TRANSFORMATION?

DRIVE PRINCIPLE #5

YOUR TRUE DESIRES, NOT YOUR INTERPRETATIONS, GUIDE YOUR ACTIONS

THIS PRINCIPLE HIGHLIGHTS THE SIGNIFICANCE OF RECOGNIZING AND REMOVING THE STORIES YOU CREATE ABOUT A GIVEN SITUATION THAT TAKE AWAY FROM WHAT ACTUALLY HAPPENED. REMOVING THESE CREATIVE STORIES ALLOW YOU TO ACT BASED ON THE TRUTH AND WHAT YOU DESIRE.

HOW DO YOU WANT TO CELEBRATE NOW OR WHEN YOU HAVE ACCOMPLISHED YOUR ORIGINALLY STATED GOALS?

WHAT ARE YOU HAPPY WITH THAT YOU ARE CURRENTLY DOING?

HOW COULD YOU ENHANCE WHAT YOU ARE HAPPY WITH USING AGILE PRINCIPLES?

WHAT IS ONE THING YOU WOULD CHANGE ABOUT YOUR CO-WORKERS OR YOUR MANAGER?

WHAT WOULD HAPPEN ONCE THIS CHANGE OCCURRED?

HOW WELL DID YOU DO IN ACHIEVING THE GOALS YOU INITIALLY ESTABLISHED AT THE BEGINNING OF THE WORKBOOK?

DRIVE PRINCIPLE #6

FOCUS ON COMMITMENTS OVER RESULTS

THIS PRINCIPLE ENABLES YOU TO RELEASE THE NEED TO CONTROL OUTCOMES. IT IS SIGNIFICANT TO ACKNOWLEDGE THE OUTCOMES TO GROW WITH THE MAIN FOCUS BEING ON THE ANALYSIS OF EFFORT PUT INTO THE OUTCOME. TRUE MASTERY FOCUSES ON THE COMMITMENT INSTEAD OF TRYING TO CONTROL THE FINAL RESULTS.

WHAT AREA(S) WOULD YOU LIKE TO FOCUS ON FURTHER?

For ELI Assessments and agile coaching services please contact us at acs.chrislewis@gmail.com or www.energydrivenagile.com.

Recommended Further Reading:
Choice Theory by William Glasser
The 4 Agreements by Don Miguel Ruiz
The Last Word on Power by Tracy Goss
Energy Leadership by Bruce D. Schneider

DRIVE PRINCIPLE #7

THE TRUTH IS GREATER THAN YOUR TRUTH

THIS PRINCIPLE PROMOTES TRUE COLLABORATION AND CREATIVITY. UNDERSTANDING THIS PRINCIPLE WILL ALLOW YOU TO QUESTION ALL YOUR BELIEFS, ASSUMPTIONS, INNER CRITIQUES, AND INTERPRETATIONS THAT ARE TELLING YOU WHAT YOU DESIRE IS UNATTAINABLE.

ADDITIONAL NOTES:

ADDITIONAL NOTES:

www.ingramcontent.com/pod-product-compliance
Lightning Source LLC
Chambersburg PA
CBHW042002150426
43194CB00002B/91